MW01101839

# Changing Coastlines

First English language edition published in 1998 by
New Holland (Publishers) Ltd
London - Cape Town - Sydney - Singapore

24 Nutford Place
London W1H 6DQ
United Kingdom

80 McKenzie Street
Cape Town 8001
South Africa

3/2 Aquatic Drive
Frenchs Forest, NSW 2086
Australia

First published in 1997 in The Netherlands as
*Kustvormen - Scheidslijn tussen land en water* by
Holding B. van Dobbenburgh bv, Nieuwkoop,
The Netherlands
Written by: Judith E.M. Peeters
Translated from the Dutch by: K.M.M. Hudson-Brazenall



ISBN 1-85368-692-1

*Editorial direction*: D-Books International Publishing
*Design*: Meijster Design bv
*Cover design*: M.T. van Dobbenburgh

Reproduction by Unifoto International Pty, Ltd

Technical Production by D-Books International Publishing/Agora United Graphic Services bv

Printed and bound in Spain by Egedsa, Sabadell

# CONTENTS



---

# PHOTO CREDITS

Arkel J. van/Foto Natura, 43[t]; BrandL H./Foto Natura, 65: Buys A. 36, 39, 42, 55[b]; Daalen F. van/Foto Natura, 18; Dobbenburgh B. van, 5[t], 30[t], 33, 67, 76[b], 77, 79[t]; Dobbenburgh M.T. van, 26, 27[b], 30[b], 50, 51, 52[b]; Dreyer G./Struik, 10[b], 11, 24/ 25, 29[t]; Eelman E.J.J., 37; Ellinger D./Foto Natura, 27[t], 70; Fey T./Foto Natua, 38, 46; Harvey M./Foto Natura, 22, 68, 69; Hazelhoff F.F./Foto Natura, 5[b], 16, 31[b]; Helo Pekka, 64 66[b]; Henstra T./ Foto Natura, 13; Hoogevorst F./Foto Natura, 23, 71, 74; Iceland Air, 45; Iers verkeersbureau, 43[t]; Koolvoort W./Foto Natura, 72/73; Lemmens F./ Foto Natura, 35,; Marissen F./Foto Natura, 12 49[b]; Martens L. 79[b]; Maywald A./Foto Natura 56/57; Meer J.J.M. van der, 63; Meinderts W.A.M./Foto Natura, 19, 20[b], 32[t], 44, 52[t]; Meyvogel J./Foto Natura, 7[t]; Nooyer P.P. de/Foto Natura, 6; Peeters J.W.L. 7[b], 29[b], 34; Schulz G./Foto Natura, 59[t]; Schtte P./Foto Natura, 28, 32[b] 40/41; Schwier P.K. 42,47, 48, 60, 61; Struik 78; Struik/ Andy S., 75; Struik/Pickfort P., 59; Sweers M./ Foto Natura, 20[t], 55[t]; Tramper R./Foto Natura, 21; Tromp H./Foto Natura, 8/9; Vermeer J./Foto Natura, 15; Verwoerd P./Foto Natura, 10[t], 31[t]; Waanders E./Foto Natura, 17[t], 53[t]; Weernik, 62; Wisniewski, W./Foto Natura 58; Wolf S. de/Foto Natura, 4;

r=right, l=left, t=top, b=bottom, c=centre

# Introduction

The earth is sometimes called the 'blue planet', because more than 70% of its surface is covered with water. At the point where water meets land, all the power of the ocean is absorbed by a small stretch of land, the coast, where processes such as wave action and ebb and flood play a role in shaping and changing this narrow strip. Many types of coast can be found all over the earth, from steep cliffs hewn from hard rock to the flat, muddy, mangrove swamps of the tropics and the artificial coastlines created by man to protect the low-lying regions of the earth from the destructive influences of the sea. The climate and the temperature of the sea water also influence the appearance of the various coastlines of the earth and sea creatures, too, leave their mark on the coasts. This book will examine all these facets of the processes that shape our coasts.

# 1 Ebb and Flood

As one of Newton's laws describes, bodies attract each other and whilst all bodies do so, the effect is usually so small that we cannot observe it. However, if the bodies are large enough, then the effect of attraction can be observed. The sun, the moon and the earth are such massive bodies that they influence each other through their gravitational pulls and despite the distance of the sun and the moon from the earth, they are still able to influence the earth, through their gravitational pull. The sun's gravitational pull is only about half of that of the moon, due to the sun's greater distance from the earth. The moon draws the earth towards it and tries to change the shape of the earth. However, given that the earth is quite solid, the bedrock that forms the continents of the earth cannot be easily changed, but on the other hand, the waters of the oceans are

*Ripples on a Texel beach in the Netherlands. The water here flowed from left to right as it ebbed. As the water dropped away these ripples were formed and left on the beach.*

*Rising tide on a beach in South Africa.*

*A first flood wave breaks through into a channel on a very gently sloped beach, where the area left dry at ebb is very wide.*

The great difference between ebb and flood was cleverly used here on the coast of Brittany, where a castle was built on a rock off the coast. At high water the fort became difficult to reach. The coast near St. Malo in Brittany, France.

Mud worms on a dry mud flat near Sligo, Ireland. Mud worms eat the sand, extract the organisms from the sand and excrete the filtered sand leaving worm casts on the mud.

much more mobile and are therefore pulled towards the moon, creating bulges on the surface of the oceans. These bulges are the oceanic tides: when a bulge approaches the coast, then a flood tide occurs; if a dip forms along the coast then it is an ebb tide. The moon's gravitational pull on the earth creates two of these bulges on the earth; one on the side of the earth closest to the moon, and one on the opposite side of the earth. The earth rotates on its own axis once every 24 hours, so that, as it turns, it passes under the tidal bulges twice a day. However, since the moon turns around the earth approximately once every 28 days, the position of the moon, with respect to the earth, changes slightly each day, causing the exact time of the ebb and flood tides to change a little every day.

The friction of the water on the bottom of the oceans causes the tides to act like a brake on the rotation of the earth, but this is a very slow process and calculations have shown that in the last 300 years the day, or rather the natural day, has become 0.002 seconds longer. This seems to be a minimal difference, but placed on the geological time scale the difference is considerable. Since the

◁Tongue-shaped ripples in Banc d'Arguin in Mauritania. This is the most important tidal region of Africa. It is affected by both the cold Canaries current and the warm Guinean current. Many tropical plants and animals reach their northernmost limits here, whilst many European and Asian species reach their most southerly limit. This kind of ripple arises through faster current speeds in the water.

*Mud flats laid bare by the ebb tide in Brittany, France. In Brittany the difference between ebb and flood can be as much as 14 metres. The first tidal power generating station in the world was opened in Brittany in 1966, providing a quarter of a million people with electricity.*

*On the rocks around Brittany, all kinds of seaweeds and grasses grow which have to be resistant to drying out at ebb.The plants not only dry out because of the low tide but because the salt in the seawater draws water from the plants at low tide. In addition they have to be strong enough to stand up to the battering of the waves.*

creation of the earth, about 4.5 billion years ago, the natural day has lengthened by about 3 hours. At some point in the distant future the earth will stop spinning on its own axis entirely, leaving constant day on one side and constant night on the opposite side of the earth, just as on the moon. In our daily lives, however, the expanding day length goes unnoticed, but what we do notice is that almost everywhere on earth is, to a greater or lesser extent, involved with tides; ebb and flood tides come and go.

Whilst the gravitational pull exerted by the sun on the oceans is small, it can be observed if the pull of the sun happens to be exerted in the same direction as

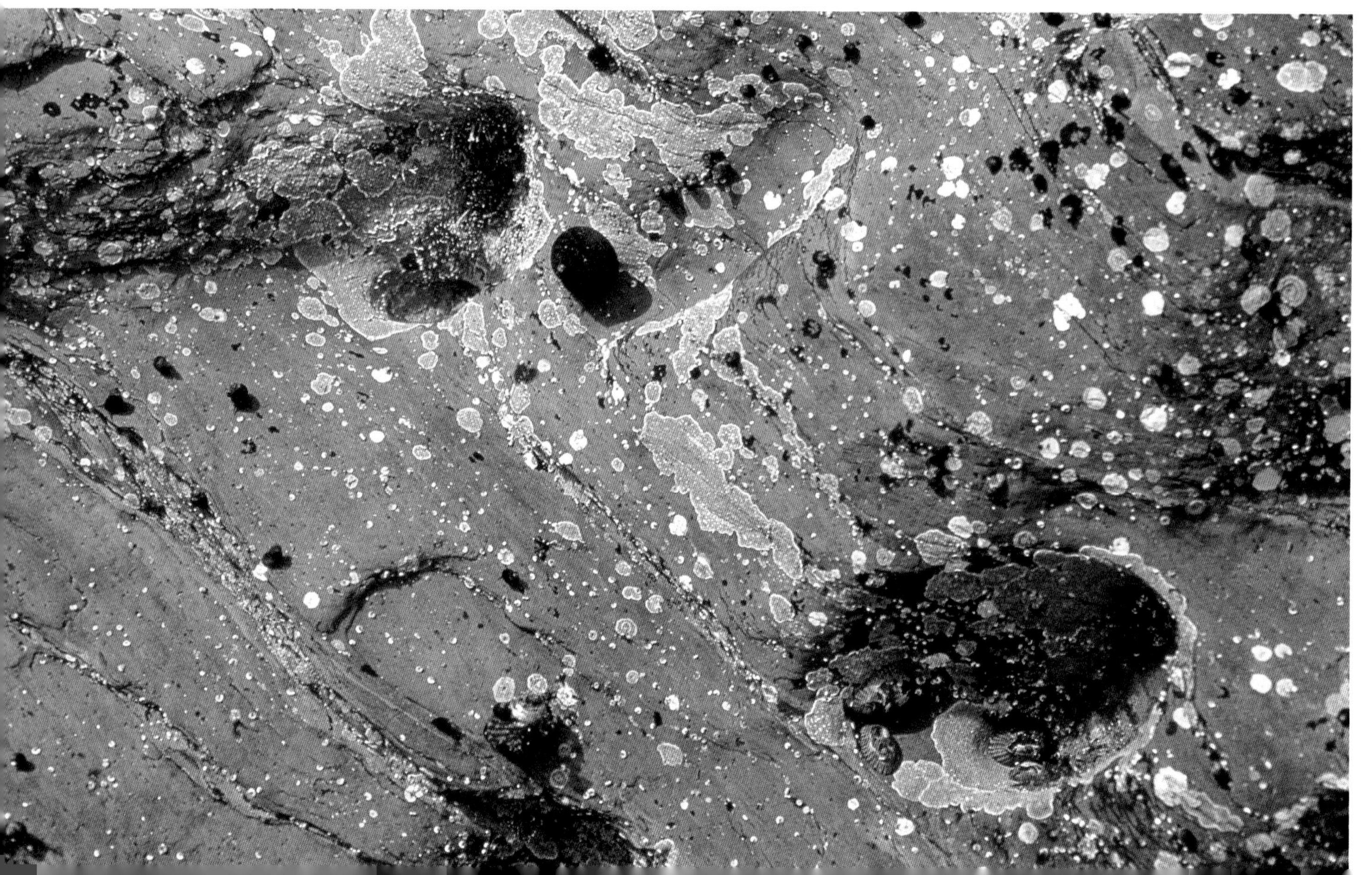

the moon's gravitational pull. This amplifies the tidal action, increasing the ocean bulges, and causing a spring tide to occur, which is higher than normal flood tides - a phenomenon that takes place once every 14–75 days. The opposite can also occur if the gravitational forces of the sun and the moon are diagonally opposite each other, then the ebb tide will be lower than normal, causing a neap tide. The difference between the normal tide and the spring or neap tide can be as much as 20% of the normal tide, nevertheless, the height of the tide not only depends on the position of the sun and the moon, but also on the geomorphology of the place where the tide occurs. The term geomorphology refers to the shape of the landscape: the size of the sea or ocean, the depth of the water at a particular place and the appearance of the seabed, for example whether there are undersea mountains or not. When we use the term, geomorphology of the coastline, we also mean specifically the coast's appearance and shape. Last but not least, the prevailing meteorological conditions at the time of a tide, such as wind force and direction, also determine precisely how the tide cycle will appear. This combination of various

*Rocks can also be broken up by the activities of sea creatures, like sponges, that bore into the rock, making it more susceptible to wave erosion. Sea urchins, too, break the rocks by boring holes into them. Storms River, South Africa.*

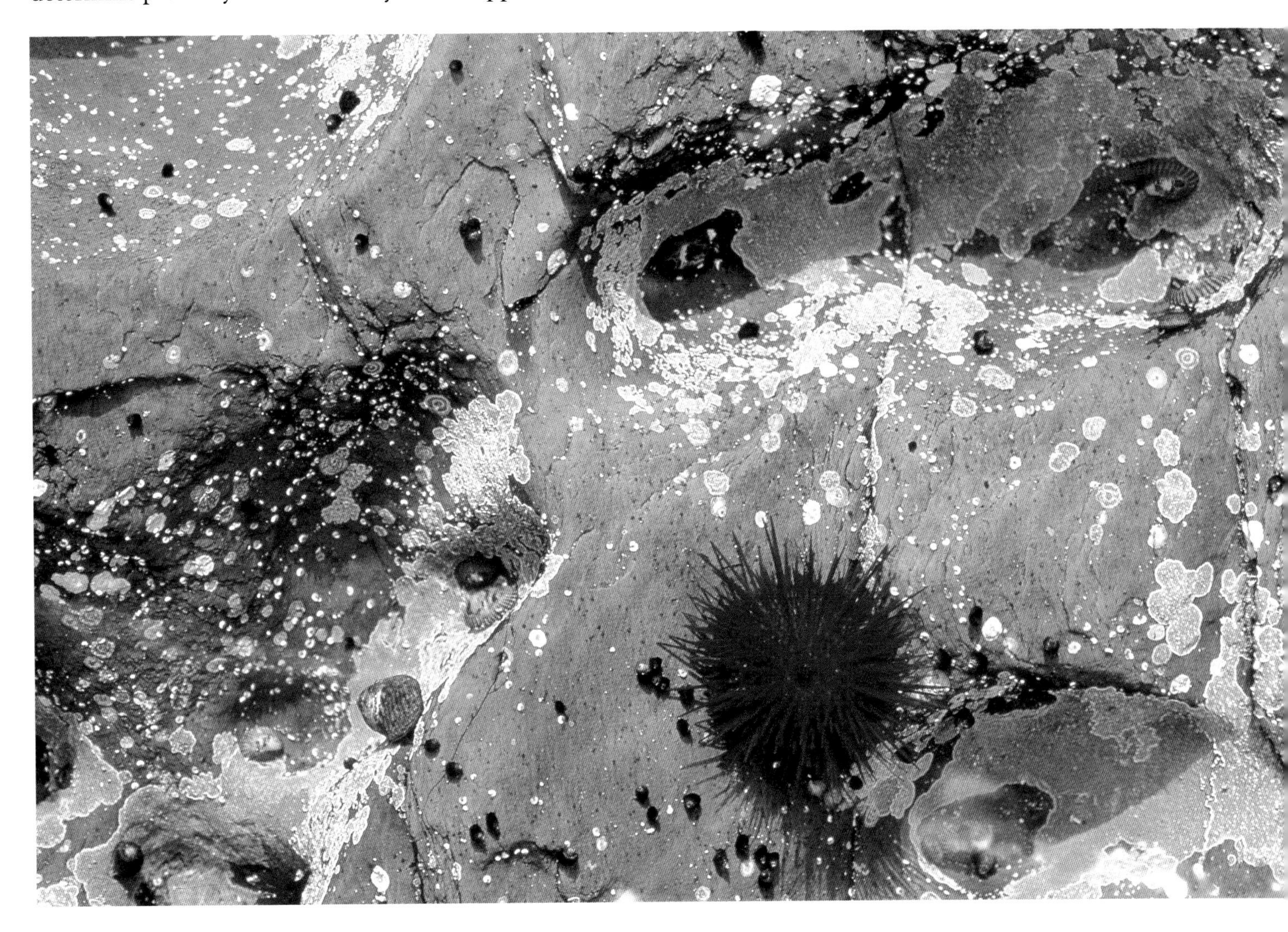

factors may result in a tidal cycle that varies from the norm of twice-daily flood and twice-daily ebb, as, for instance, along the Antarctic coasts where there is a once-daily tide cycle, producing flood and ebb only once each day or, like the mixed tides that occur in the Pacific Ocean, producing ebb and flood tides which in each period of 24 hours are of different heights. The exact cycle of the tides is not, in itself, important in determining the shape of the coastal morphology, but the difference between ebb and flood is of far greater

*Water streams at ebb back to the sea through the lower sections of the mud flats. The lower-lying sections are thereby deepened even further by the erosive action of the water. Water does not flow back to the sea like a blanket, but gathers and forms still wider and wider channels. These channels do not start being very wide but can finish up as wide as a river. In the three different zones of the mud flat the channels have different names. In the salt marshes they are called creeks; in the area between the high and low water lines they are called tidal gullies and below the low water line they are called channels. They cut through the sand banks. The tidal area is a very dynamic environment. This can be seen from the many abandoned meanders, which are filled with fine sand and mud because the water is left standing in them. These pools are often full of sea creatures left behind on the ebb tide.*

significance. In almost fully enclosed seas, like the Mediterranean and the Black Sea, the differences are minimal, whilst along open coasts the differences between ebb and flood can be as much as six metres. In north-western Europe the tidal range is very large, as demonstrated along the Channel coast of Brittany where the difference between ebb and flood can, for example, be 14 metres, because the water is funnelled into the English Channel. In St. Malo, in Brittany, they have taken advantage of the large tidal range to build a tidal electricity generator to harness the energy produced by the movement of so much water back and forth. A substantial length of the Canadian coast also experiences a large tidal range with the largest differences between ebb and flood occurring in the Bay of Fundy in north-west Canada: here the difference is nearly 16 metres.

In addition to the geology of the coastline, the type of tide cycle and the difference between ebb and flood also determine the shape or morphology of the coast. The type of tide cycle determines the period between flood and ebb tides, thus in a semi-diurnal cycle, with ebb and flood twice a day, the time between low and high tides is, naturally, shorter than in a once-daily cycle. The longer the time between high and low tides, the greater the influence of weathering and biological activity that is exerted on the exposed seabed;

moreover, with a shorter time between ebb and flood, the water flows faster, increasing the erosive power of the water. This effect is especially large in semi-enclosed bays where the water is forced through a small entrance channel. The difference between ebb and flood determines the vertical distance over which currents and waves can exert their influence on the coast and as a result, the gradient or slope of the coast thus determines the area of land that falls dry at ebb and is covered at flood. The more gentle the gradient is, and therefore the flatter the land, the larger the area is that falls dry at ebb.

The deltas of the great European rivers, like the Rhine and the Meuse, lie in north-western Europe, flowing into the North Sea, one of the regions in the world where the difference between ebb and flood can be as much as six metres. Deltas are very flat areas: the gradient is very gentle, which allows the influence of the tide at times to be quite considerable. A large proportion of the coast around these deltas is protected by rows of dunes, a phenomenon whose origins will be examined in detail in the next chapter. In the past, a rise in sea levels resulting from the melting of icecaps and glaciers at the end of the last ice age, meant that the sea broke through these rows of dunes in the area between the northern Netherlands and south-west Denmark, thus forming the Wadden Sea, the largest salt water marsh in Europe. The tide cycle in this region is semi-diurnal, with ebb and flood tides occurring twice a day. The vast area of the Wadden Sea is almost emptied at ebb tide leaving huge mud flats which become flooded and fall dry twice a day.

The mud flat area can be divided into three zones, with the lowest zone, below the low water line, being an area that is always covered with water; the intermediate zone, lying between the high and low water lines, which is inundated with water and falls dry twice daily and the highest zone, lying above the high water line, and is only occasionally submerged by water. As one can imagine, huge quantities of water stream in and out of the region as a result of these tidal movements. The ground in the region consists of sand, which can

*A channel in the Verdronken Land (Drunken Land) of Saaftinge, The Netherlands. The Verdronken Land ensued from the All Saints Flood of 1570, when the area was partially flooded. Later, in 1584, it was completely flooded by man to prevent the advance of the Spanish army during the Eighty Years War. Afterwards the land was never reclaimed from the sea and now it functions as flood containment to protect the port of Antwerp in Belgium.*

be picked up and transported by the current and then deposited elsewhere, causing the water to shape and form the sand in several different ways. In water that flows in one direction for a reasonably long time at a certain speed, flow ripples are created, which on the upstream side, have a long slope and on the downstream side, a shorter, steeper slope and are therefore asymmetrical in shape.The ripples show the direction of the current through their shape. Geologists use this characteristic to determine the direction of flow of the water that first formed these sediments. Once formed, the ripples become fixed in the sediment as a result of the sediment particles sticking to themselves and turning to rock through the process of petrification, which preserves the flow ripples for millions of years. Where sedimentary structures remain preserved, geologists are able to establish old deposition patterns from these ripples. However, sedimentary structures are not always preserved, as the sun may destroy them by drying up the sand, allowing the structures to crumble. The sand may then be transported further by the wind and wading birds, too, may disturb the structures. The size of the ripples depends on the size of the grains of sand in which the ripples are formed and on the strength of the water current: in general, the finer the sand, the higher the ripples. In strong currents too, the ripples will be larger, but when the current is too strong, then no ripples are able to form. In a tidal region, where currents flow in various directions because the water first flows into the area and then back out again, the flowing water will carry sediment in various directions. The stronger the current flows, the larger the sediment particles that can be transported. Reversing the direction of flow in a tidal region as the tide changes means that the water actually stands still for a short time. This period is called slack water. When the tide turns from flood to ebb it is at this point in time, that the greatest area of the tidal region is under water. Since water that stands still can no longer carry sediment, this results in the sediment sinking, depositing a layer of clay over the ripples previously formed in the sand. Clay is the collective name for the finest particles of sediment that sink last; that is, only when the water stands still. As the water starts flowing back to the open sea, new ripples form which are laid down over the clay. The leaf-shaped structure of the clay particles makes it difficult for the water to pick them up, leaving them undisturbed when the water starts to flow back to the open sea. The new ripples that are then deposited on top of the clay, lie in the opposite direction to the first layer of ripples. The structure, which arises in this manner, since it resembles the bones of a fish, is called herringbone cross-bedding by the sedimentologists, and is a pattern of deposition that is very typical of tidal regions. In the zone below the low water line, clay layers are also found that formed during the transition from ebb to flood. Through the continuous deposition of sand and clay the mud flats come to lie higher and higher and they can even lie higher than the average flood tide line, which means that they lie above water during the average high tide. These mud flats are often then colonised by halophytic plants, these being plants that have developed special defence systems against salt water.

# 2 Dune Landscapes

When there is a constant and sufficient supply of sand to the coast, sand dunes will form, which are created as a result of the sand on the beach drying up at low water and being carried along by the wind. Dry sand allows all kinds of structures to be formed in the sand by the wind. So-called wind ripples are formed, which look very similar to flow ripples, but with one major difference: wind ripples only form when the wind is strong enough and there is sufficient fine sand. The various ways that the sand is transported by the wind depend on the size of the grains of sand since sand grains smaller than 0.025 millimetres are able to remain in suspension in the air and sand grains that have a diameter between 0.01 millimetres and 0.5 millimetres, seem to bounce along the surface. The threshold for transportation of sand is reduced by this bouncing action, for once it has been picked up by the wind, sand that lands on the

*Whirlwind on a dune. Dunes are formed by the wind carrying and then depositing sand in one place.*

*◁Structures which are shaped by the wind in sand, are often laid one on top of each other, as on this dune on the beach at Terschelling, The Netherlands, on which ripples created by the wind have also formed.*

*On the beach on Terschelling, The Netherlands, the dry sand is blown together by the wind forming dunes. The wet sand sticks together and is not moved around by the wind.*

*Wind speed is highly variable, around obstacles like plants, so that the sand being carried by the wind can find itself in calm air causing it to drop and no longer be carried. Here on Terschelling in the Netherlands, small dunes can be seen forming around the vegetation.*

ground is deflected back into the air by the grains of sand lying on the ground. The collision with the grains on the ground provides them with enough energy to stay in the air, whilst at the same time it is also possible for grains lying on the ground to be loosened, as a result of the collision with grains landing on the ground, and then lifted into the air where they are carried along. Larger grains between 0.5 and 1 millimetre may simply be rolled along by the wind. When there is an onshore wind of sufficient strength then the sand is blown inland out of the reach of the water. Once there, piles of sand soon mount up, especially around obstacles like large plants and trees. If the sand settles on the ground then it is much more difficult to transport it any further. Obstacles can

*Dunes can, once they have been colonised by primary vegetation, become even bigger, after which even larger plants can establish themselves. De Zilk, The Netherlands.*

be almost anything: rubbish that is lying on the beach for example, but also plants that are growing higher up the beach.
The beach is an unfriendly environment for most plants, for it is, on the whole, very dry with few nutrients in the soil. It is comparable to a desert environment, for here too, the differences in temperature can be very large. The plants also run the risk of being buried under migrating sand or inundated by salt water. The plants that grow on the shoreline dunes, have all adapted to these factors. They belong to the primary dune vegetation. The shoreline dunes are those dunes that come into view first from sea. A common type of grass found on the shoreline dunes and the landward dunes is marram grass, which has leaves that can be rolled up if the plants threaten to dehydrate. The stomas, where water normally evaporates from, can be closed to protect the plant from drying out. Marram grass also has the ability to grow quickly up through the sand if buried, its spear-shaped leaves pushing the sand away. These qualities make marram grass highly suited to the maritime environment of dunes and it is often specially planted to help in the formation of dunes, holding the sand with its roots so that it can no longer shift. Dunes can grow very rapidly: in 24 hours, tons of sand can be shifted by the wind.
The shoreline dunes are straight lines of dunes that have formed directly on the beach. The landward region of the dune zone, however, does not consist of evenly spaced dunes. Why is this? Given that dunes are very unstable land forms, whenever they are no longer stabilised by vegetation or other obstacles, the sand is able to shift again. The vegetation in the dunes is very susceptible to

*Sand has a critical angle of stability. If the angle at which the sand is lying is greater than this critical angle, then the slope becomes unstable and the sand will start to shift. In wet sand this angle is greater than in dry sand and the slopes are steeper. The chances of the sand shifting is then greater if the sand dries up. Schiermonnikoog, The Netherlands.*

*In the Slowinsky National Park in Poland there are shifting dunes which move at a rate of ten metres per year. On their way they engulf the woodlands that they encounter. Decades later, once the dune has shifted on, the dead skeletons of trees that were buried under the advancing dune reappear.*

*The internal structure is clearly visible in a section of dune that has been washed away somewhere along the North Sea coast. During a single storm tons of sand can disappear back into the sea.*

disturbance. Man for example, walking in the dunes or trying to farm or letting animals graze there, will disturb the vegetation and allow the dunes to shift again. Due to the turbulence in the air and the various obstacles, different sorts of dunes can then be formed because the sand is in motion. The exception is wet or damp sand, which shifts less easily than dry sand, so that in the damp dune valleys the sand will therefore be less likely to shift.
The highest dune in Europe, the Dune du Pilat, is 2.7 kilometres long, 500 metres wide and 114 metres high and lies near Arcachon in France. In the Les Landes region to the south of Bordeaux where this dune can be found, the shifting sand has been halted successfully by planting pine woods, but they have been less successful in halting the Dune de Pilat: it is still shifting inland to this day.
Dunes, of course, are not only formed, but they can also disappear again through the action of the sea. During a single storm, tons of sand can disappear back into the sea, which can have major consequences for low-lying

*To protect the beach from erosion, groynes are erected along the beach to break the waves. The waves have lost much of their erosive action by the time they reach the beach. The beach by Schoorl, The Netherlands.*

*Most dunes are not to be found along the coast but in the great deserts of the world, where an almost endless quantity of dry sand is available for the formation of dunes. In Namibia, the Namibian Desert lies directly adjacent to the Atlantic Ocean.*

areas lying behind with a shoreline with a shallow slope. The Wadden Sea in north-western Europe was formed in this manner when the sea broke through the lines of dunes, which stood along the line of the present-day Wadden Islands, inundating the area lying behind them. The Hondsbossche Sea Wall, which is some five kilometres long, lies between Petten and Groet in north-west Netherlands and was built to protect the low-lying areas from the sea and to replace the line of dunes washed away in the St Elizabeth Flood in 1421. Another way to protect coasts from coastal erosion or dune breakthrough is to build piers or groynes that stand at right angles to the shoreline. This reduces the force and therefore the erosive action of the waves on the beach and dune line, and also by damping the wave action reduces the transportation of sand along the coast.

# 3 Waves

Waves are formed by the action of the wind pushing across the water and transferring the wind energy to the water, with each wind speed being associated with its own characteristic waves. Waves can be characterised in terms of three dimensions: the height of the waves (the amplitude of the wave), the distance between two successive wave tops (the wave length) and the time between the passage of two successive wave tops (the period). These three characteristics of waves are determined by the speed of the wind, the length of time that the wind has been blowing for, and the distance the wind has already travelled over open water: thus the stronger the wind the higher the waves. The height of the waves is, however, fixed to a certain maximum, because of the development of white crests at a particular height. Since the formation of white crests requires energy, which is also derived from the wind, it cannot therefore put energy into building height into the waves. With a slight breeze, the wavelets that form as soon as the wind starts to blow also rapidly fall away

*Waves on the ocean. The ocean loses all its energy in the narrow transitional zone between land and water, which is shaped by that energy.*

*The surf in Camps Bay, Cape Town, South Africa. The southern tip of South Africa is notorious for its violent winter storms.*

when the wind drops, whilst when the wind is stronger, more than three kilometres per hour, waves are formed that no longer fall away immediately when the wind drops. These more persistent waves are then able to leave the storm area where they were formed and roll further across the ocean. A complicated pattern of waves is created on the ocean by waves from many different storm areas running into one another and it consists of a combination of the waves formed in far-off storm areas and waves that have formed under the influence of local winds. As long as the waves roll through deep water they remain unaffected by the ocean floor, but when the water depth reduces to less than half the wave length, as is the case when waves approach the shore, then the wave action is also affected by friction on the seabed. The wave is then slowed by the seabed, but the waves that are in shallower water are overtaken from the sea by waves which still have the speed that the waves had on the open sea. This has the effect of shortening the wave length, i.e. the distance between successive wave tops, so that if the wave length and the speed are reduced, then there is more energy left to put into the height. A wave that comes into shallow water therefore continues to gain height, but because the height of a wave is fixed to a certain maximum the wave will break then at some point. In addition, as there is less friction on the upper

*Waves approaching the beach in South Africa. They touch the sea bed with their base and are slowed down. As the top is still travelling faster, the top overtakes the base and the wave then breaks.*

*Waves flowing back to the sea also drag back with them, some of the material that they have tossed onto the beach when they turned. Cousin in the Seychelles.*

*Trees washed up on the beach on Vancouver Island, Canada.*

side of the wave, it is not slowed down as much as the underside, also causing the wave to break. This is called the surf: the wave flows out over the slope of the beach until all the energy is used up and then the wave streams back, partly over the sand, partly through the spaces between the grains of sand; this action causes the wave to take sand back to sea that, only a short time before, had been thrown up on the beach. The beach is therefore the place where all the energy of the waves combines and is also lost, and this can have destructive consequences.

When waves break on the beach then actual movement of the water takes place, which is not present in the rolling action of waves on the ocean: there the water simply turns in circles under the wave. Through the movement of water in the surf, all sorts of materials can be transported like sand and

▷Waves breaking in Storms River, South Africa.

When strong onshore winds blow, the waves are higher because the waves tops are swept up even more. Kemner Dunes, South Pier, Ijmuiden.

pebbles, with the waves throwing them repeatedly onto the beach and then dragging them away producing attractive, round pebbles. Pebble beaches are always steeper than sandy beaches because the water drains away quicker between the pebbles than in fine sand and since pebbles are also heavier, once they are thrown up on the beach, they are less easily drawn back into the sea, because the waves expend the last of their energy on the beach. A beach consisting of pebbles can actually have a slope of some 20 degrees whilst beaches consisting of fine sand generally have a slope of only a few degrees. Often the expenditure of wave energy is not so gradual as in the surf of a gently

▷Raindrops only form around minute particles floating in the atmosphere. One of the inexhaustible sources of these nuclei is salt from sea water that is released as sea spray in the surf. When waves break, microscopic particles are produced of which only the salt is left over once the water evaporates. Cahill Island, Co. Mayo, Ireland.

sloping beach that is formed in unconsolidated sediments. Waves can, on their way to the beach, meet obstacles that cause them to expend their energy all at once. This is how trees that lie in the surf can be stripped of their bark and sanded smooth by the material transported by the waves. This also happens on rocky coasts: the rolling waves then break on the hard rocks, so that even if a rocky coast looks massive and impenetrable, there are almost always hairline cracks in the stone of which the rocks are made. In these hairline cracks there is air and when a wave breaks on the rocks, the air in these hairline cracks is compressed with great force. A normal Atlantic wave breaking on rock, for example, in the winter exerts a force on the coast equivalent to 10,000

*One can tell from the surf how deep the water is. Waves only break if they come into shallower water and their bases touch the sea bed. Along this coastline in the vicinity of Hot Bay, South Africa, the shallowing of the ocean can be observed from the surf line.*

*Waves transport materials like sand and rocks, which they continually throw onto the beach and then drag back into the sea. This action rounds off both sand and rocks and smooths them into pebbles. Grinaví, Iceland.*

kilograms per square metre and during storms this force is even greater. When the wave falls back, the air then expands very rapidly, breaking fragments off the rock in a similar fashion to the effect of an explosion and leaving holes or fissures in the rocks. Sometimes a fissure in the rocks will break through and appears at the top of the cliff on the surface as a hole close to the edge of the cliffs. When a wave then runs in, an enormous amount of water is suddenly pushed by the force of the wave into a small opening, causing a sort of fountain

Sea creatures that live in the tidal zone on a rocky coast must adapt to the daily cycle of ebb and flood, like these sea anemones that withdraw into an envelope. Sea anemones are not plants but animals that catch small sea creatures with their tentacles. Brittany, France.

Seaweeds that grow on cliffs have to be able to resist the forces exercised by the wave action. To do so the seaweeds that grow in these places have developed all sorts of mechanisms, like rapid regeneration, so that sections of plant that have broken off are quickly replaced with new parts. Brittany, France.

*These tree stumps on Vancouver Island, Canada (above) and Oregon Beach, USA (below), have been stripped bare through the action of the weather.*

to spout out at the top of the cliff. Geographers call this a blowhole. In the course of time these blowholes become larger and larger due to the continuing erosive action of the waves. Sea caves, where the roof collapses forming a tunnel between sea level surface and the top of the cliffs, are also called blowholes, although the water does not necessarily spout like a fountain. If a wave rolls into such a blowhole then a gust of wind can be felt, which is pushed ahead of the wave.
The tallest wave, created through the force of the wind, that has ever been

witnessed, was seen on the night of 6–7 February 1933 on the Pacific Ocean. The wave was moving west to east from the Philippines in the direction of California and was estimated to be a colossal 34 metres high. Waves caused by undersea earthquakes can, however, be even higher and are known by their Japanese name, tsunami. Tsunamis arise from the movement of the ocean floor as a result of an undersea earthquake and they usually begin as very shallow waves with a height of 1 metre and a wave length between 100 and 700 kilometres. Their starting speed is between 500 and 900 kilometres per hour, but once they reach shallow water then their speed slows and their height consequently increases. The highest tsunami ever seen occurred on 24 April 1771, and measured an astounding 85 metres, and was powerful enough to carry a piece of coral weighing 700 tons the distance of 2.5 kilometres inland. After they form, tsunamis reverberate back and forth across the Pacific Ocean

*The great force exerted by breaking waves on the coast squeezes the air in the hairline cracks in the rocks. When the wave falls back the air suddenly re-expands, just like in an explosion. Existing hairline cracks are thus made wider and wider.*

*Crashing waves along the coast at Kalk Bay Harbour, South Africa.*

Due to the great force with which water is forced through the tunnel of a blowhole it becomes wider and wider. When a blowhole enlarges it no longer spouts in normal weather when waves run in but a gust of wind can be felt. Blowhole near Downpatrick Head, Co. Mayo, Ireland.

because they are so enormous, often taking days to die away. In 1960, it was not only the villages and towns in the area around the epicentre of the earthquake in Chile that were destroyed, but 22 hours later and some 1,7000 kilometres away in Japan, the coastal town of Honshu was also destroyed by the tsunami created by this earthquake. For a number of days the tidal observation centre in Hawaii was still recording differences in tides as a result of the tidal wave reverberating back and forth across the Pacific Ocean.

Waves can also be caused by volcanic eruptions as demonstrated by the explosive volcanic eruption of Krakatau in Indonesia in 1885.This eruption was so violent that it created a bang which was heard in Australia 2,000 kilometres away, and resulted in a hole opening up in the sea bed that filled up at an enormous speed, creating a tidal wave 30 metres high. On Java and Sumatra more than 36,000 people died as a result of this tidal wave and it even caused floods in Australia. Before its energy had fully dissipated the tidal wave created by the Krakatau eruption had travelled three times around the earth.

In storms the air pressure in the centre of the low pressure area can fall by as much as 100 mbar which results in air pressure being 15 % lower than normal. The low air pressure allows the water of the ocean to be sucked up to one metre high, forming a storm wave which, especially if it occurs in combination with high tide and an enclosed sea, can have destructive consequences.

▷Ecuador: waves pounding powerfully on the coast. If the water is squeezed with great force through narrow clefts, then it spouts out. These are called blowholes.

# 4 Cliffs

Cliffs are steep, rocky precipices that rise up out of the sea. The highest cliffs in the world are 1,010 metres high and lie at an angle of 55 degrees and can be found on the north-west coast of Molokai, one of the Hawaiian islands. The 400 metres high Conachair Cliffs on St Kilda, to the west of the Hebrides, are the highest cliffs in Europe.

The angle that the cliffs lie at depends on the type of stone and how many fissures and hairline cracks criss-cross the rock, as well as the form and degree of erosion acting on the cliff. Along any coast the most important form of erosion is wave action but waves only have an erosive action if they meet the coast in water that is shallow enough to allow the waves to break. Where the water at the base of the cliffs is too deep for waves to break then wave erosion will not occur. Cliffs like these are known as plunging cliffs, and they can often be found in drowned glacial valleys, along geological faults and around

*If waves are given long enough to eat away at the rock, then tunnels can eventually develop in rocky promontories along the coast. These are called sea arches. The coast of south Victoria, Australia.*

volcanoes. Cliffs with shallow water directly below them will be attacked by the waves breaking at their foot. Waves only rarely approach a coastline at a right angle: usually the movement of the waves lies at an angle to the coast. The waves that are first to arrive in the shallow water, are slowed down by the seabed and then overtaken by the waves that arrive later, resulting in the waves bending so that they still reach the shoreline in a straight line.
Sandy coasts are, on the whole, easily moulded by the waves and when viewed from the sea appear as a straight line; this is not true of rocky coasts which are, by their very nature, not straight. Whenever waves approach a coastline that is not straight, almost the same effect occurs as when waves approach the

*When sea arches are forming the nature of the rock is very important. The rock must not be too hard, for then the waves cannot break it up, but the rock must also not be too soft, because then it crumbles too fast and the arches collapse. Banda Jissa, Oman.*

coastline at an angle. The waves are bent, around rocky headlands, in the direction of the rocky point because the sea bed is shallow. This means the waves slow down in the shallow water and the following waves, being faster, then overtake them, causing the waves to bend towards the promontory and break there. More waves, therefore, break on a promontory, length for length, than break on an equivalent piece of straight coast so that the promontory has to withstand more force from the waves. In the bays behind the promontories

Sea arches, once formed, are still subject to the influence of the pounding waves and will continue to be eroded. The Gap, Albany, south-west Australia.

The span of a sea arch will gradually, through the ceaseless action of the waves, become thinner and thinner until it collapses. Great Ocean Road, Australia.

◁An attractive example of a sea arch on the Azores, Portugal.

*Along the Oregon coast (USA), it is often misty. Warm sea water evaporates here along the coast and forms a thick cloud layer that results in heavy rainfall, especially in the coastal regions of Oregon. Despite its northerly latitude, Oregon's climate is remarkably mild because of this warm current.*

*When the span of a sea arch collapses, this leaves a single rocky promontory called a sea stack. A new sea arch has formed in this sea stack off the coast along Great Ocean Road, Australia.*

the opposite occurs: there are fewer waves breaking, causing more deposition. The sea is thus actually busy straightening the coastline.

The promontories are more exposed to erosion because more waves break on them. Each wave that breaks, compresses the air in the hairline cracks in the bedrock, thus forming holes in the rock. If two of these holes which are on opposite sides of a promontory make contact with each other, they then form a tunnel allowing the wave action thereafter to continue undiminished, widening the tunnel. This is how a sea arch is formed, but their creation depends on the kind of bedrock: it has to be solid and resistant to erosion. The hardness of the bedrock also determines how large such an arch becomes: if the stone is hard, such an arch can become very large; if the stone is soft then the arch will break up and fall down, just leaving a pillar, called a sea stack, in the sea, a short distance from the shoreline. Not only are the promontories eroded by the sea but also the relatively straight cliffs, which are undermined by the waves breaking on the base of the cliff, leave overhanging rocks which may then collapse, causing the cliffs to recede. The speed with which this occurs depends on the degree to which the rock is eroded, given that in relatively soft rock this will happen faster than in hard, solid bedrock. One famous example

◁The continuing erosive action of the waves can leave sea stacks even more remote from the coast as the coast retreats and the sea stack gets smaller. Sea stack near Downpatrick Head, Co. Mayo, Ireland.

▽Sea arches have problems forming in rocks which are too crumbly. The Cliffs of Moher, Co. Clare, Ireland are one example. These cliffs are 230 metres high.

of this are the chalk cliffs of the Channel. Since Roman times, some 2,000 years ago, the cliffs on the English side have receded between three and five kilometres causing whole villages to disappear into the sea. The cliffs of the Channel are composed of the chalk skeletons of animals and plants that lived in the seas millions of years ago which, when they died, sank to the bottom, where they formed a thick layer. The cliffs are composed of layers of this chalk interspersed with layers of sand, which forms a very soft combination, which is relatively rapidly eroded by the waves of the Channel. At the foot of such a cliff, a smooth, scoured plateau, the abrasion platform, is left at the depth where the waves still have an erosive action. This occurs particularly in areas where the difference between ebb and flood is not very large and the wave action is concentrated on a limited zone of the cliff. In this situation the cliff may retreat from the actions of the waves, leaving the waves breaking only on the abrasion platform. Often this platform is covered with pebbles, which, because they are

*△A double sea arch in the rock Hvítserkur that lies off the coast of Iceland. According to an old legend it is a troll that changed to stone at sunrise.*

*◁All rocky vertical outcrops that form off the coast are called sea stacks even though they have never had an arch structure as a starting phase. This phase can be skipped if the stone is too crumbly for a sea arch to form. This sea stack off the coast of Scotland shows that no arch has formed because the stone is too crumbly.*

*The sea stack, Breanan Mor, lies off the Cliffs of Moher, Co. Clare, Ireland. This sea stack is 70 metres high, and pales in comparison with the cliffs that tower to 230 metres.*

moved around by the waves, erode the platform even further. As the platform is eroded the wave action no longer erodes the base of the cliff and the cliff then becomes a fossil or abandoned cliff.

Although it is the dominant erosive process in most places on earth, wave action is actually not the only one responsible for eroding cliffs. Weathering and mass movements also contribute, especially in those areas where wave action is slight; other forms of erosion are weathering by salt, frost weathering, solution and biological weathering and in areas where wave energy is low, erosion by sea creatures plays a dominant role. Molluscs, sponges and sea urchins bore holes in the rock, causing it to break up; a phenomenon that appears particularly in the tropics. Salt and frost have the same effect on stone, for when salt is precipitated on the rock and water freezes in the rock, they both expand, forcing fissures in the rock to open wider.

The gradient or slope of the cliff is partly determined by the dominant erosion process, wave action, as well as the type of bedrock that the cliff is made of, the geological layering of the rock and how many fissures run through the bedrock.

Rocks only dissolve when they consist of chalk, and chalk dissolves much more rapidly in the tropics than in temperate climates because the temperature is higher and more carbon monoxide is present. The more carbon monoxide that is dissolved in the water, the more chalk that can also be dissolved in the water. This carbon monoxide comes from the breakdown of organic material, that is abundantly available in the tropics and rapidly decomposed in the high temperatures and humidity. In chalk plateaus, conical depressions that are called dolines, are created by the solution of chalk. In the tropics these dolines expand quickly in width and depth, becoming so large so that eventually island mountains form between the dolines. These mountains are the only remnants

*Pancake rocks on the west coast of South Island, New Zealand. The formation looks like stacks of pancakes because of the build-up of thin layers of chalk.*

▷Rocks off the Scottish coast.

of the initial plateau and are very steep-sided and full of caves. This form of karst is called tower or conical karst.
In Ha Long Bay in North Vietnam, the bedrock was originally a Palaeozoic chalk plateau which has, in the course of time, mostly disappeared through the solution of the chalk, now leaving only conical mountains. The chalk which appears here on the surface is the same as the chalk from which the famous conical mountains in Guilin, China are formed. In Ha Long Bay the sea has now drowned this landscape leaving the mountains to become islands with steep cliffs. According to Vietnamese mythology, the shapes seen in the rocks

*△▷The Twelve Apostles are a famous rock formation off the coast of Victoria, south Australia. There are actually only ten pillars. No one knows whether there were ever twelve or whether the other two have been engulfed by the sea, or whether it was really only a case of poetic licence when the rocks were named.*

were formed by a mythological dragon; Ha Long means the place where the dragon descended. In many places in the world the local population have dreamt up myths to explain natural phenomena that they observed. This happened as well in the Waitaki District of South Island, New Zealand.
Here, in the coastal cliffs, the Moeraki, stones can be found; these are mixtures of chalk, silicon, aluminium and iron formed around a nucleus of chalk. The Maori, the native people of New Zealand, have their own explanation for the boulders that they found on the beach, which is that, according to them, these stones are food baskets that have been washed up after the canoe, Arai-tiuru, was shipwrecked after a long voyage across the ocean of Kiwa, otherwise known as the Pacific. The canoe still lies on its side a short distance from the beach: this is the reef lying offshore. The hills, that reach down to the sea, are

▷*Three-master in the harbour of Reykjavik, Iceland.*

the passengers of the canoe who reached the shore uninjured but who were surprised by the sunrise. The sunrise had a special significance for the Maori, given that it represented the end of the old day and the beginning of the new. Nowadays only the largest stones are left on the beach: the smaller ones have all been removed as souvenirs by visitors.

In Northern Ireland there are also some spectacular cliffs that go by the name of the Giant's Causeway and they, too, have been given a mythological significance. The cliffs here consists of basalt columns formed by the volcanoes of the Tertiary Period (about 50 million years ago) which were active in this region. The columns were formed by lava cooling slowly and solidifying

*Rocks (66 metres high) rising out of the sea near Vik, off the southern coast of Iceland. They are claimed to be the remains of a three-mast ship that was dragged into the sea by trolls. However, they were surprised by the sunrise and they and the ship were turned to stone.*

*Monument of a Viking ship, Reykjavik, Iceland.*

whereby shrinkage clefts formed that extended lengthwise. Most of them are hexagonal prisms about six metres tall with a cross-section of about 50 centimetres, but there are also pentagonal and decagonal columns. The largest columns are 12 metres tall. The cliffs look like street paving because they are stacked in such a regular pattern, which is also why they were so named. According to tradition, the Causeway was built by the Irish giant Finn MacCool, who built a road over the sea to his enemy, Finn Gall, in order to challenge him to a duel. Finn Gall lived on the island of Staffa that lies 120 kilometres from the coast of Northern Ireland and one day when Finn MacCool was lying resting, Finn Gall took the opportunity to cross over to Ireland. He saw Finn MacCool sleeping, but thought that this was the son of Finn MacCool and was shocked by his size. He fled back to Staffa, destroying the road built by Finn MacCool as he went, leaving the Giant's Causeway as the only remnant.

*△◁Undercut cliffs with an abrasion platform formed by the abrasive action of the waves. Mainland Orkney, Scotland.*

*◁Waves undercut the cliffs that then crumble. The crumbled rocks are then rounded by the wave action.*

*Waves have a limited influence in water. The base of the wave is the maximum depth at which the wave can erode. The erosion is caused by material on the sea bed being moved about by the wave and scouring the sea bed. Thus abrasion platforms are created that fall dry at low tide like this one at Cap Frehel, Brittany, France.*

*Rock formation off the coast of Brittany, France.*

*Rugged coast consisting of soft and hard layers along the Great Ocean Road, south Victoria, Australia.*

The island of Staffa, and in particular, Fingal's Cave, also consist to a large extent of the same basalt columns that can be seen at the Giant's Causeway. Cliffs are favoured breeding places for colony-nesting seabirds because there are only a few suitable places allowing their preference to nest close to the sea in order to be able to feed easily. Birds also nest in colonies in places where suitable nesting sites seem to be plentiful, for the larger the number of birds in the colony the smaller the chance that an individual will be attacked. Penguins breed in colonies for this reason although there seem to be sufficient breeding sites. It is often safer to breed on cliffs because the breeding birds and their young are less easily accessible to birds of prey or other predators like foxes. Puffins (Fratercula arctica) prefer, for example, to breed in very large colonies on steep cliffs. To protect their young from predators they dig burrows anything from 20 centimetres to 2 metres into the cliff or use cracks in the

*◁Some bedrocks are so unconsolidated that they form screes. If these screes are steep they too are called cliffs. These cliffs are a good example of the undermining of the cliff by the waves.*

*The Giant's Causeway in Northern Ireland.*
*A rock formation consisting of columnar basalt. According to legend these are the remains of a street built by giants.*

*Chalk cliffs are not only easily eroded by the waves, but they also dissolve readily in water, causing them to retreat even faster. Mecklenburg, Germany.*

*When cliffs retreat, they can come to stand out of reach of the waves through the formation of an abrasion platform.They are then called fossil cliffs. Hangklip, Cape Province, South Africa.*

rocks, but from the moment that the young appear from the burrow they are highly vulnerable to attacks by birds of prey and other predators. Sea bird colonies that nest on level ground lose more young than those on steep cliffs for this reason, but man is also responsible for losses as large colonies of seabirds are often the target of egg collectors. The birds may also be 'harvested' by the local people, as on the Faeroe Islands where puffins have been caught with nets on sticks ever since anyone can remember.

*◁The German island, Rügen, consists of soft chalk that dissolves in the sea water. Sometimes the water is white with chalk. These chalk cliffs, deposited more than 65 million years ago during the Cretaceous period, were covered by glacial moraines during the last ice age. As the cliffs retreat the stony morainic overburden falls onto the beach.*

*In the bedrock that forms the cliffs at Waitaki on South Island, New Zealand mixtures of aluminium, chalk, silicon and iron can be found.*

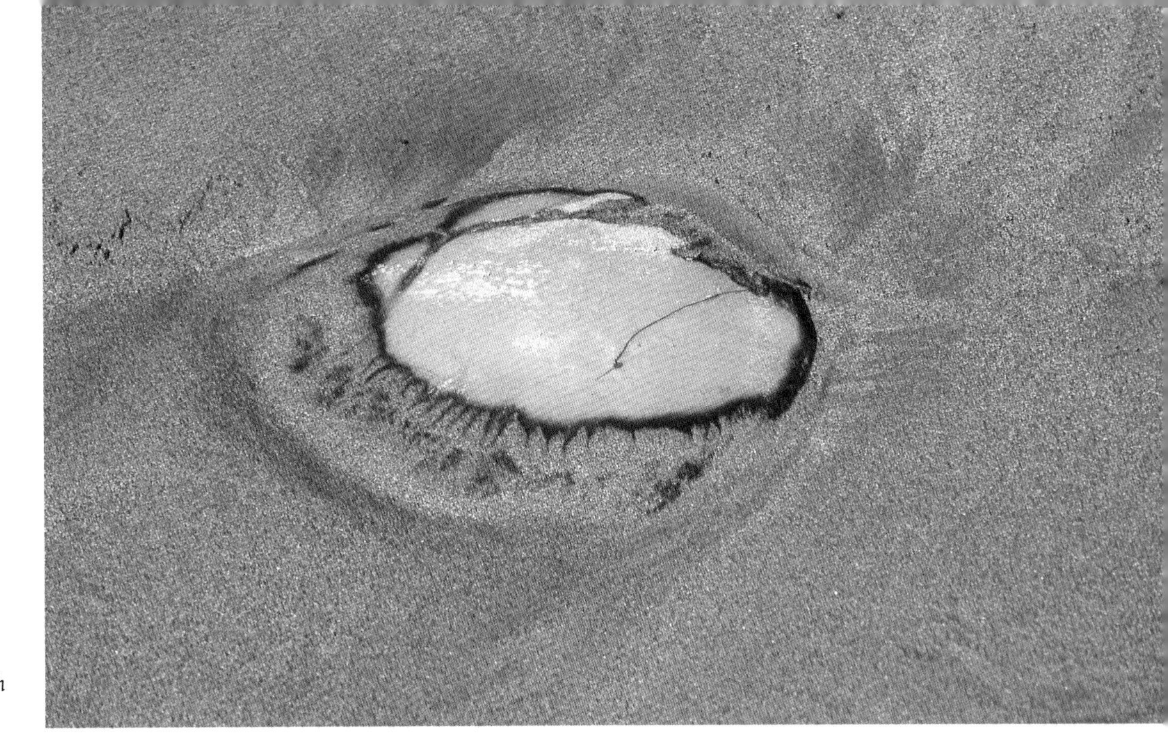

When the cliff retreats the boulders fall onto the beach. Waitaki, South Island, New Zealand.

Nowadays only the largest of the boulders remain as the rest have all been removed as souvenirs. Waitaki, South Island, New Zealand.

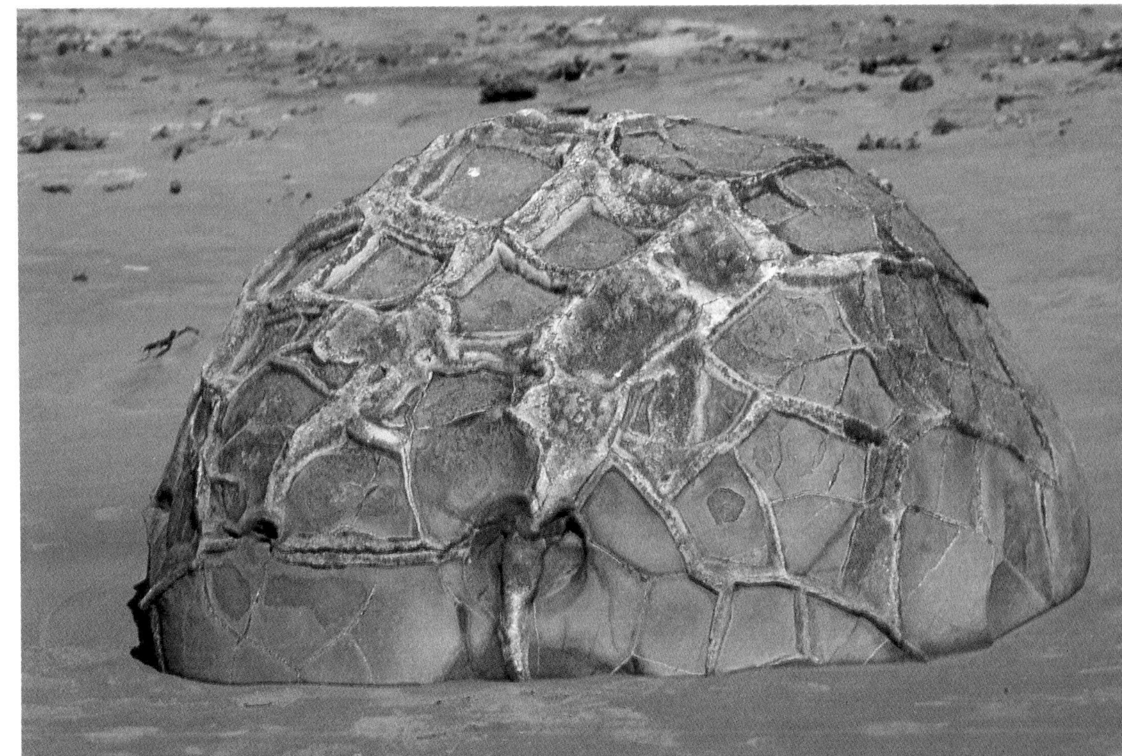

Ball-shaped mixture of iron, aluminium, silicon and chalk on the beach at Waitaki, South Island, New Zealand.

# 5 Fiords

Fiords are deep, spectacular sea inlets with very steep sides, which form some 30.7% of the entire coastline of the world. They are found particularly in the higher latitudes, along the coasts of Norway, Sweden, Denmark, Canada, the United States, Chile and New Zealand. How are they formed?
Scientists working at the beginning of the nineteenth century had already discovered that the climate of the world has not always been the same. They found deposits in places where there was no ice, which were very similar to deposits that occur in places where there are glaciers and icecaps today and drew the conclusion that there must have been ice there in the past. This is called the uniformity principle, which means that the way in which deposits or sediments are laid down remained the same throughout the whole history of the earth.
There seem to have been periods in the earth's history in which places, where temperate or even tropical conditions prevail at present, have been so cold in the past that ice could form and remain for longer periods. This was apparent from the sort and amount of the deposits that were laid down, as well as their extent, so that by examining the sediments closely, scientists have been able to establish where ice formed in the past. It would seem that during the last ice age 29% of the total land surface area was covered with ice, an area of some 44 million square kilometres. At present, only 10% of the earth's surface is covered with ice, and of that Antarctica constitutes 84%. The warm and cold periods have interchanged throughout the earth's history and geologists have been able to establish that in the last 1.2 million years there have been four cold

*Fiords are created by the abrasive action of glaciers. Fiords can be much deeper than river valleys as the erosive effect of the glacier continues underwater due to the weight of the ice. Glaciers are formed when more snow falls in the winter than melts away in the summer.*

*Fiords are typically U-shaped: steep valley sides and a fairly flat bottom, which is not often visible because the fiords are filled with water.*

periods, called glacials, and the same number of warm periods, known as interglacials. The glacials lasted much longer, with the last glacial in our latitudes only ending 10,000 years ago. During a glacial, ice caps form first in the high mountains, and subsequently the ice flows out along the existing valleys to the lower-lying regions, much like the glaciers nowadays in the Alps and the Rockies. During a glacial these high mountain glaciers expand and then also cover the lower-lying areas. Ice does not only leave deposits but can have an abrasive action, that is to say, the ice breaks off the material that it flows over. The valleys over which the ice pushes its way down, are worn away by the ice and reshaped. Valleys that are formed by rivers typically are V-shaped but the abrasive action of the ice changes this to a U-shape as can clearly be seen in, for example, Alpine valleys. When glaciers are being formed, the precipitation that lands on the glacier is slowly transformed into ice. This precipitation comes from the oceans and becomes locked up in the glaciers and icecaps which implies that the sea level during the ice ages was much lower than in the warm periods, perhaps a fall in sea level of as much as 150 metres.

In Norway and Sweden, at the beginning of the last glacial, there was a mountainous region from where the glaciers could grow. During the ice age the area was covered in a blanket of ice up to 3,000 metres thick and it is this ice mass which has, to a great extent, defined the landscape and coasts of

Norway and Sweden. Ice, like water, flows downhill and under the influence of the gigantic weight of the ice a glacier will also move very slowly downhill. The ice follows the existing valleys, which are carved out into a U-shape and are further eroded, which is the same way that fiords were formed. Most fiords were carved out by glaciers that finished up in the sea, which left the fiord about 300 metres or more deep, with a shallow threshold lying at the mouth of the fiord. The threshold is the remnants of the terminal moraine left behind by the glacier. The Sogne Fiord in Norway is 1,300 metres in depth whilst the mouth of the fiord is only 150 metres deep. The reason that fiords are so deep is because the glacier does not stop its abrasive action when it dips into the sea. At certain depths of water, depending on the thickness of the ice cap, it will float and its erosive action will cease. Thus a 300 metre thick glacier will only begin to float if the water is more than 270 metres deep. The threshold at the fiord mouth marks the point at which the glacier lost its erosive power. Moreover, glaciers create terminal moraines by dumping the debris that they have carried with them at their nose. This is, however, not the sole explanation for the depth of the fiords because, when the ice caps melted at the end of the last glacial, the sea level rose by about 100 metres.
Fiords often form good harbours, not only because they are so deep, but also because of the steep sides, due to their typical U-shape, so large ships with a deep draft are able to come alongside. However, this also implies that there is

*When fiords have very steep sides then there is not much habitable land left. Through their steep sides the fiords are good natural harbours. Geiranger Fiord, western fiord region, Norway.*

*◁Fiords extend deep inland. The Sønderestrøm Fiord in Greenland is 160 kilometres long. The largest fiord in the world, the Sogne Fiord in Norway, reaches as far as 200 kilometres inland.*

*Varanger Fiord, Norway.*

*Not all fiords have steep sides. Some have formed in less mountainous regions, so that the fiord sides are less distinct. Narvik, Norway.*

little inhabitable ground in the valley of a fiord.
One of the many spectacular phenomena associated with fiords are the waterfalls that plunge into it. These occur because the side valleys, that open into a fiord, lie high above the level of the fiord floor or the sea level. The reason for this is that the side glaciers that fed into the main glacier only eroded down to the level of the glacier that used to fill the main valley . When the glacier melted away, these hanging U-shaped valleys were left.

# 6 Tropical Coasts

Coral islands, with their azure blue waters, white sands and waving palms trees, belong to the coastlines of the world that most appeal to our imaginations. Often they lie in remote peaceful regions of the tropical seas, making them even more attractive. Corals do not grow everywhere in the tropical seas of the world. So why does coral grow in one place in the tropics and not in another? The reason why coral islands only occur in the tropics can be found in the coral itself, which consists of innumerable polyps, that themselves consist of a bony skeleton covered in a soft multi-coloured tissue. Polyps themselves are about five millimetres in diameter; a reef is therefore made up of billions of them. There are some 350 sorts of polyps. Fish, sea stars and other spiny sea creatures can also be found on the reef, as well as micro-organisms which also excrete chalk, that then sinks to the bottom and cements the polyps together. The coral polyps only grow in water with an average temperature of 24°C, therefore corals mainly occur along the eastern seaboards of the continents. As cold water frequently wells up along the west coasts of the earth's continents, the required average temperature of 24°C cannot be achieved. Coral begins to grow if there is a sufficiently strong, stony sea bed available to which the polyps can attach themselves. This sea bed must, however, not lie at too great a depth, otherwise the polyps will get too little light. The maximum depth at which coral will grow is 90 metres below sea level, but the ideal depth is 20 metres below sea level. Such a stony base can be provided by underwater volcanoes, which are tall enough to meet the needs of

*The natural cover of palm trees binds the sand, protecting the island from erosion. Riau Archipelago, Indonesia.*

*A coral island on the Great Barrier Reef in Australia. The colour difference in the water reveals the extent of the coral reef. The surf line marks the edge of the reef and the point where the sea bed falls away to deeper waters.*

the coral. Corals do not, however, occur around every tropical island nor on every continental east coast because they do not thrive where there is a lot of sediment and they also require water that has a high salt content. If too much fresh water from rivers dilutes the salt water it lowers the salt content making it too low for corals to grow.

In the Atlantic Ocean, between 30°N and 30°S, in what is essentially considered to be the tropics, few corals grow. This is due to a heavy sediment load carried into the Atlantic Ocean by major rivers like the Mississippi, the Niger and the Amazon, which prevent sufficient light penetrating the depths to allow coral to grow. The coral would also be covered in layers of sediment carried down by these rivers. Most corals occur in the eastern section of the Pacific Ocean.

Coral reefs can be divided into three groups: fringe reefs, barrier reefs and atolls. This distinction is made on the basis of how the reef develops. Reefs develop around islands in tropical waters and these islands are often volcanoes. The reef begins by forming around the volcanic island, creating a circular reef around it, which, in this state, is called a fringe reef. A fringe reef has formed in this way around Molokai, in Hawaii. As volcanoes that stand in the sea gradually sink under their own weight back below sea level this causes the surface area of the island to shrink slowly. Given the very slow rate at which the volcano sinks, the growth of the reef manages to keep pace with the relative rise in sea level resulting from the slow sinking of the seabed on which it stands. The reef retains its original circumference, whilst the surface of the island becomes smaller and smaller, so that between the reef and island, a steadily increasing gap is created, which fills with coral sand. A shallow lagoon develops between the island and the reef and at this stage the reef is known as a barrier reef. An example of a barrier reef is to be found off the island of Moorea, one of the Society Islands. The final stage that a reef can achieve is that of an atoll. An atoll is created when the volcanic island disappears completely under the sea leaving behind only a ring-shaped reef, enclosing a shallow lagoon. Bikini Atoll in the Marshall Islands is a very well-known atoll. Islands of coral sand can form within an atoll, which are very sensitive to disturbance. If the vegetation on such an island is removed for agricultural purposes, then the sand is no longer retained and can be washed away in a heavy rain shower. The coral of the reef may then become choked as a result of being covered in a layer of sediment washed off the island into the lagoon, and this may then lead to the death of the reef.

Low-lying regions in the tropics, that are subject to tidal action, are not places where mud flats and salt marshes develop, but they are home to mangrove swamps. Mangrove swamps are woods that are specially adapted to life in a salt environment but which, like the temperate salt marshes, are only underwater for some of the time. The trees have developed special air roots which provide the roots with air when the swamp is underwater.

Mangrove swamps, or rather their roots, hold the sediments in place. Mangrove swamps can be found in Indonesia, northern Australia, the Amazon delta and in the Niger delta in Africa. They contain a very rich variety of plants and vary in composition from site to site, depending, in part, on the salt content of the water, the tidal action and the composition of the soil. More than 80 mangrove plant species can be distinguished in mangrove swamps. The roots of the mangroves provide a sheltered nursery for many marine animals but it is a very fragile ecosystem because mangroves grow in a very dynamic and, therefore, very unstable environment.

On many tropical beaches, beach rock can be found, which consists of sand that is cemented together by calcium carbonate that has precipitated out because of the high rate of evaporation in the tropics. The calcium carbonate originates in the ground water in the sand and precipitates out as the water

*Around this coral island off Queensland, Australia, a barrier reef has formed. This is the second stage in the creation of coral reefs, according to Charles Darwin. First of all, a fringe reef forms, that becomes a barrier reef when the island begins to sink back into the sea.*
*Finally, when the island disappears below sea level again, what is left is an atoll.*

Coral islands are surrounded by a shallow lagoon between the actual island and the outer edge of the reef. Tropical fish and other sea creatures are found in abundance in the lagoon. Anse source d'Argent, La Digue, Seychelles.

Tropical islands are typically covered in palm trees. When they are chopped down, the sand that the island is built of, washes away and chokes the coral, which in turn gets too little light and dies. Anse source d'Argent, La Digue, Seychelles.

Rock formation on the beach at Anse source d'Argent, La Digue, Seychelles.
La Digue is an island with a granite core of which this rock formation is also formed. Coral reefs have formed around the island.

evaporates. If water, in which a certain substance is dissolved, then evaporates, at some point this substance will precipitate out of the water. At higher latitudes chalk also occurs on beaches, but there, the rate of evaporation is not so high that it is able to precipitate out again. Once beach rock has formed, it is very resistant to erosion and it protects the beach against the erosive action of the waves.

*Another form of tropical coastline is the mangrove swamp that grows in the flat tidal regions. These are the mud flats of the tropics. The roots of the trees stabilise the mud and thus protect the coastline. Mangrove trees have air roots to provide their roots with oxygen.*

*Weathered, salted white, this tree stump lies on the beach of Poivre West, Seychelles.*

# 7 Famous Capes

Capes do not only form important landmarks on the coast but also often in the history of man. A cape is actually nothing more or less than a headland or land tongue in the sea, but it is precisely their historical value which makes capes so special. The first rounding of a cape often has a symbolic significance because it marks the attainment of a certain milestone. The most famous capes in the world were named at the time of the great voyages of exploration which took place at the end of the Middle Ages and during the Renaissance. A list of the famous capes of the world would naturally include the Cape of Good Hope in South Africa and Cape Horn in Chile, but also the North Cape in Norway, which forms the most northerly point of mainland Europe as well as Cabo da Rocha, near Lisbon in Portugal which is the most westerly point in Europe. One of the most symbolic capes is deservedly called the Cape of Good Hope in

*Cape Horn, Chile.*

South Africa, which lies at the point where the Atlantic Ocean meets the Indian Ocean at the place traditionally known as the 'Gateway to India'. In contrast to what many people think, the Cape of Good Hope is not the most southern point of the African continent, that honour falling to Cape Agulhas, a couple of hundred kilometres further to the south-east. In the Middle Ages, Europe was supplied with spices, seasonings and silk overland via the Silk Route and the trade with the Far East was in the hands of the Arabs. At the end of the 14th century the Turks appeared in the Near East making trading more difficult and the goods more and more expensive. Faced with a growing demand for spices, seasonings and silk in Europe, any European country to find its own route to the Far East would be ensured of a golden future. The Portuguese explorer,

*Table Mountain, Cape Town, South Africa, seen from Blouberg.*

Bartholomeo Diaz, was sent by his monarch to attempt to discover the sea route to the east and he was the first European to round the Cape of Good Hope, even if he himself did not realise that he had done so, for it happened in a raging storm. The Cape was, therefore, first named the Storm Cape because of this. The area around the Cape of Good Hope is renowned as an area where the weather can suddenly change. Diaz himself did not discover the route to the east as he was forced to return to Portugal by a mutiny on board. King João II of Portugal christened the Cape, the Cape of Good Hope, because the

discovery had increased the chance of finding a route to the east. The Portuguese had been busy exploring the coast of Africa since 1433, a seemingly endless task, but with the discovery of the Cape of Good Hope there was suddenly hope of finding the route to the east.
This route to the east was discovered by Vasco da Gama in 1497, another Portuguese explorer who had accompanied Diaz on his journey in 1486. The first staging post to be founded at the Cape of Good Hope was established on 16 April 1652 for the East India Company by a Dutchman named Jan Riebeeck. A permanent Dutch colony was set up in the following years, which was later seized by the English and it was only in 1910 that the then independent colony was added to the Union of South Africa. The Cape of Good Hope is not only

◁*The weather can suddenly change in Cape Province: huge storms can sweep in, unexpectedly uprooting trees and changing rivers into raging torrents. The sea then pounds against the coast. This is how the cape got its first name: Storm Cape. Cape Peninsula, South Africa.*

renowned for its illustrious past, but has also gained notoriety in recent times because of Robben Island. This island, that lies a few kilometres offshore, is the place where Nelson Mandela, the symbolic leader of the anti-apartheid movement, was incarcerated for many years. Table Mountain, towering above Capetown is also a very well-known landmark. Somewhat less well-known is the fact that the peninsula, which forms the Cape of Good Hope, has a completely separate flora, with plant species that only grow here and nowhere else in the world.
Following the discovery of the Cape of Good Hope, the trading empire of the Portuguese was rapidly established in the Far East and the Portuguese monarch was therefore not disposed to finance an expedition proposed by Ferdinand Magellan, namely, finding the westerly route to India. This expedition would only cost money and do nothing to enhance Portugal's position. Magellan

*Cape Province in South Africa is the world's smallest, completely separate flora district. Nowhere else on earth has so many different species of plant in such a small area. Cape of Good Hope, South Africa.*

◁*The most northerly penguin colony in the world lies in Cape Province, near Boulders Beach,*

*Simonstown, South Africa. The first European to round the Cape of Good Hope was Bartholemeo Diaz who was searching for a route to the East on the orders of João II of Portugal. He was blown round the cape during one of the many notorious storms that blow there. He finally landed in Mussel Bay, 200 kilometres further up the coast. A mutiny forced him to return to Portugal. Cape of Good Hope, South Africa.*

persevered and finally set sail with Spanish money to search for the westerly route to India. America had been discovered some years previously, but he believed that there had to be a route to the Southern Ocean, as the Pacific Ocean was then known. In the wide estuary of the Rio de la Plata, near Buenos Aires in Argentina, they thought they had discovered the route, but that was not to be, and it was only when they were much further to the south that the coast opened up again and Magellan decided to take a chance and sail into the rather unattractive straits, which to this day carry his name: the Straits of Magellan. The most southerly point of the continent actually lies a little further south and this was rounded by a European for the first time in 1616, a Dutch seaman by the name of Willem Cornelisz Schouten, who named the cape after his birthplace, Hoorn, in North Holland. The cape consists of a 300 metre high rocky promontory and is 8 kilometres long. The westerly winds make the voyage around the cape a risky adventure and many sailing ships have been shipwrecked on the rocks here, not only because of the storms that blew here,

*Cape Town and Lion's Head seen from Table Mountain. Cape Town was founded in 1652 by a Dutchman travelling to the east, Jan van Riebeeck. In the distance lies Robben Island, where Nelson Mandela was imprisoned for many years. South Africa.*

*The Cape of Good Hope has been for centuries the gateway to the east for trading missions from Europe. The sailors knew that once they had rounded the Cape of Good Hope, then there was a good chance that they would reach India.*

but because of the treacherous mists and rain that obstruct the view. Most ships, which sailed the westerly route to the Far East, followed the more sheltered route through the Straits of Magellan. Since the opening of the Panama Canal in 1914, the trade route around the southerly point of the continent has fallen into disuse and Cape Horn has become one of the most remote and inhospitable regions of the earth.